Daniel Deleanu

Notes in
Logosophistic Medicine

LogoStar Press

Toronto

ISBN: 978-1-105-42580-6

Printed in the US.

Introductory Study

Towards a Logosophistic Diagnosis
In Medicine

Logosophistic diagnosis is a term I have used in the study of contemporary medical sciences in order to name a process by which diseases appear to be *logoi* (patterns stemming from an archetypal Logos – a sort of hypothetical "super" Reason which is the etiological matrix of all medical conditions that shift and complicate in medical reasoning when analyzed in light of the conjectures and gaps they reveal within themselves).

Subjects relevant to logosophistic diagnosis include pathophysiology and the ways that *disease connotation* is constructed by medics and understood by patients. This process is based on analogy – (from the Greek *ana-logos*), an anaphoric regression towards the hypothetical Logos of the disease. Logosophistic diagnosis is a different method of medically diagnosing a patient, which involves discovering, recognizing, and understanding the underlying—both explicit and implicit—assumptions, ideas, and frameworks that form the basis for medical investigation, diagnosis, treatment, and prognosis. Medical logosophistic diagnosis has a philosophic foundation in my logosophism, Heidegger's notion of *Destruktion* and Sartre's concept of the Other.

The problem of defining logosophistic diagnosis

The term *logosophistic diagnosis* in the context of modern medicine is resistant to formal definition. Its central concern is to de-construct and then re-construct traditional medical pathology.

When asked to define logosophistic diagnosis, I have no simple and formal response to this question. All my answers are just attempts to have it out with this redoubtable question. There is a great amount of bewilderment as to what kind of thing logosophistic diagnosis is — whether it is a school of medicine (it is certainly not so in the singular), or just a new method of differential diagnosis (it has been reduced to this by various attempts to define it formally) — and establishing what ability to accord to a particular endeavour to draw up its boundaries.

Surveying medical logosophistic diagnosis, one is confronted with a puzzlingly diverse range of arguments. These include claims that logosophistic diagnosis could sort out the Western tradition of medicine in its entirety by emphasizing and discrediting unjustified privileges accorded to various approaches, some of whom have been proven to be erroneous. On the other hand, some critics claim that logosophistic diagnosis could be a precarious form of investigation that wishes the utter destruction of the Western medical and ethical values. While there is no doubt that principal approaches associated with logosophistic diagnosis may be fruitful in their medical investigations, the place of differential diagnosis in logosophistic approaches complicates matters considerably.

Logosophistic diagnosis utilizes an idiosyncratic (sometimes, in fact, mimetic) style with a massive amount of allusion across many corners of the Western canon of medicine. Critics assert that when one takes the time to logosophistically analyze diseases, one discovers it was not worth the effort because it allegedly brings nothing else but a refreshed form of differential diagnosis. However, logosophistic diagnosis is ultimately and literally a comparative medical diagnosis of the symptoms of a condition. It is characterized by connotation and is, in fact, a direct attack on any possibility of denotation, because if the medic alone creates denotation in labelling a disease, there can be no truth; one person's denotation is equal to another's, so there is no possibility of consistent discourse leading to actionable conclusions in diagnosis, treatment, and prevention. Logosophistic diagnosis leads nowhere and denotes nothing, since it is not its purpose do that. Unlike differential diagnosis, which uses a deductive technique, logosophistic diagnosis utilizes a methodology grounded on both deduction and induction.

An apophatic approach to logosophistic diagnosis

It is probably easier to define what logosophistic diagnosis *is not* than what it is. Logosophistic diagnosis is not an examination, nor an evaluation, a process, an act, or an operation. Further, logosophistic diagnosis is not, properly speaking, a synonym for investigation. Rather, it is a specific kind of analytically reading symptoms.

Logosophistic diagnosis is an analytical method which is in fact much closer to the original meaning of the word "analysis" itself, which etymologically means "to undo" — a virtual synonym for "to dismantle." If anything is dismantled in an analytical reading, it is not the medical information, but the claim to undeniable dominance of one mode of signifying over another. A dismantling reading of a medical symptomology, for example, is a reading which analyzes the medical specificity of a variation of itself.

Some critics of this approach claim logosophistic diagnosis amounts to little more than medical relativism. However, it is not truly the abandonment of all pathophysiological meaning, but attempts to demonstrate that Western medical knowledge has not satisfied its quest for a "transcendental" signifier (the medical Logos) that will give meaning to all other medical signs. Logosophistic diagnosis is not an enclosure in a pathogenetic vacuum, but an openness to the Other (the Other disease) and an attempt to discover, not the place, but the non-place which would be that Other disease and the *logoi* of its variations.

Approaching a definition of logosophistic diagnosis

Part of the difficulty in defining logosophistic diagnosis in medicine arises from the fact that logosophistic analysis cannot escape itself. The word itself is subject to the linguistic limitations and effects which it purports in its own definition. One should not view logosophistic diagnosis as a concept standing outside of its textual medical corpus, which can act upon all medical bodies without itself being affected. The act of defining, in this respect, logosophistic diagnosis is an attempt to finish or complete a medical investigation and taxonomize its finds; however, logosophistic diagnosis should never be viewed as a complete nosology, but rather as a continuous process, a living medical philosophy with pragmatic ends being constantly adjusted within itself.

Nonetheless, one can provide a number of rough definitions. One of them might state, for instance, that it is possible, within the textual corpus of a medical investigation to frame a question or undo assertions made in the Other textual corpus, by means of elements which exist in *any* medical textual corpus, and which frequently would be precisely structures that play off the taxonomical against the investigative. Viewed in this way, the term "logosophistic diagnosis" refers in the first instance to the way in which the "accidental" features of a medical textual corpus can be seen as unrevealing and subverting its essential elements through idiopathic mechanisms.

Whenever logosophistic diagnosis finds an idiopathic nutshell – a secure medical axiom or a Hippocratic maxim – the very idea is to crack it open and disturb its tranquility. Indeed, that is a good rule of thumb in logosophistic diagnosis. That is what logosophistic diagnosis is, in fact, all about, its very meaning and mission. One might even say that cracking idiopathic nutshells is what logosophistic diagnosis is. In establishing diagnoses, have we not run up against medical paradoxes and aporias? The intellectual paralysis and experimental impossibility of a medical aporia is, *a priori*, just what impels logosophistic diagnosis.

Some definitions could portray logosophistic diagnosis as a method, project, or school of medicine. Logosophistic diagnosis signifies, *de facto*, a project of critical medical thought whose task is to locate and take apart those concepts which serve as the axioms or rules for a certain thought, those concepts which command the unfolding of an entire outset regarding a certain medical condition. Logosophistic diagnosis is somewhat less negative than the terms "destruction" or "reversal" (of a certain medical conception); it suggests that certain foundational concepts of medicine will never be entirely eliminated. There is no simple overcoming of the chemical and physical processes occurring in the human body. Similarly, in the context of medical studies, we can define logosophistic diagnosis as a way of uncovering the questions behind the answers of a certain medical textbook or tradition.

Medical logocentrism and the critique of primary oppositions

The central concern of logosophistic diagnosis is the identification in the Western medical tradition of a "logocentrism" which holds the archetypal Logos (the hypothetical Reason which is the etiological template of all medical conditions) as a privileged, ideal, and self-present entity, from which all medical discourse and meaning are derived. This medical logocentrism is the primary target of logosophistic diagnosis.

One typical form of de-constructive/re-constructive reading is the critique of primary oppositions, or the criticism of mono-logic thought. Logosophistic diagnosis is by definition dichotomous thought, that is, it utilizes dialogic discourse. A central de-constructive argument holds that, in all the classic dualities of thought, one term is privileged or "central" over the other. The privileged, central term is the one most associated with the hypothetical medical Logos. Examples include:

- clinical medicine over medical research
- physical examination over medical history
- assessment over differential diagnosis
- treatment over prevention
- laboratory studies over imaging investigations
-etc.

In each such case, the first term is traditionally conceived as original, authentic, and superior, while the second is thought of as secondary, derivative, or even "sponging." These disjunctive oppositions, or hierarchies, and others of their form, must be de-constructed in order to re-construct them according to the requirement of each and every case.

The logosophistic diagnosis is effected in two ways. However, these oppositions cannot be simply transcended; given the thousands of years of medical history behind them, it would be devious to attempt to move directly to a domain of analysis and beyond these distinctions. So, logosophistic diagnosis attempts to compensate for these historical power imbalances, undertaking the difficult project of thinking through the taxonomical implications of questioning and presenting complications to show the eventuality of such divisions.

Loghosophistic diagnosis *de facto* involves the materialization or outbreak of a new conception. One can begin to conceive a conceptual complementary terrain away from these oppositions: the next project of logosophistic diagnosis would be to develop concepts which fall under neither one term of these oppositions nor the other. Much of the theoretical work of logosophistic diagnosis should be devoted to developing such ideas and their implications, of which the Logos (the hypothetical etiological Reason of medical conditions) may be the prototype (as it denotes neither simple unity nor simple difference of meaning).

It is necessary to analyze, to set to work, within the text of the history of medicine, as well as within the so-called medical textual corpus certain *logoi* that by analogy (*ana-Logos*, or going back to the archetypal Logos) I have called undecidables, that is, signs of simulacrum, "false" medical marks that can no longer be included within contextual opposition, resisting and dis-organizing it, without ever constituting a tertiary term, without ever leaving room for a unitary solution in the form of speculative pathological or etiological dialectics.

As can be seen in this discussion of its terms' undecidable complexity, logosophistic diagnosis requires a high level of analysis with deferred decision; a medical analyst must be willing to work with terms whose precise meaning has not been, and perhaps cannot be, established. Critics of logosophistic diagnosis find this unacceptable as methodology; many feel that, by operating medical *logoi* in this manner with unspecified terms, logosophistic diagnosis ignores the primary task of medicine, which they say is the elucidation of concepts. This criticism is a result of a fundamental difference of opinion about the nature of medicine, and is unlikely to be resolved in a simple manner.

Etiology and logosophistic diagnosis

According to de-constructive interpreters of logosophistic diagnosis, one of the centrisms of modern medicine is the distinction between the Logos and logos, with the latter historically being thought of as derivative to the former. As part of subverting the presumed dominance of logos over the Logos, one can argue that the idea of a Logos-logos dichotomy contains within it the idea of a very expansive view of etiology that subsumes both Logos and logos. In fact, there is nothing outside of the etiology of a disease. That is, etiology is thought of not merely as linear thinking derived from the Logos, but any form of representation, marking, or storage, including the marking of the human brain by the process of cognition or by the senses.

In a sense, logosophistic diagnosis is simply a way to read a medical configuring taxonomy; any logosophistic diagnosis has a medical con-figuration as its object and subject. This accounts for logosophistic diagnosis' broad cross-disciplinary scope. Logosophism has been applied to literature, art, mathematics, philosophy, and psychology, and any other disciplines that can be thought of as involving the act of textually configuring an epicentral Logos.

In logosophistic diagnosis, con-figuration can be thought of as unmarked, in the sense that once the con-figurations are made, the taxonomy remains in suspended vibrancy and does not change its structure. Thus, what the authors of a medical textbook say about their con-figuration of a condition does not restore its lost original *imago*: it is just another medical con-figuration building on the original, along with the building of others. In this view, when a medical author says, "You have comprehended my taxonomy perfectly," this utterance constitutes an addition to the configuring system based on the hypothetical Logos, along with what the said was understood in and about the original con-figuration, and not a restoration of the original *imago* of the condition, which is now lost. The diagnostician has an opinion, and the patience has an opinion too. Logosophistic medicine does not exclude the patient from the equation of medical diagnosis. Communication between the physician and the patient is possible not because the text has a transcendental signification, but because the physician's diagnosis contains similar con-figurations of the disease as the patient's autotelling symptomology. These con-figurations, however, are both unstable and incomplete, or even disconnected sometimes from the etiologic line conducive to the medical Logos.

Unconfigurability

Logosophistic diagnosis exists in the interval between configurability and unconfigurability. The chief exemplar of this relationship is the relationship between physiology, pathophysiology, and etiology. Idiopathy summarizes the relationship by saying that justice is the unconfigurable condition that makes logosophistic diagnosis possible. However, this relationship, especially in the case of idiopathy, is indeterminate and not a transcendental, ideal one which leads us to the con-figuration of the Logos. In fact, it would be an indeterminancy in itself, if such a thing like the medical Logos existed outside or beyond itself.

Medical taxonomy is made up of necessary medical con-figurations, while idopathy is the unconfigurable. Logosophistic symptomology belongs to the realm of the present, possible, and assessable, while traditional etiology belongs to the realm of the absent, impossible, and inestimable. Logosophistic diagnosis bridges the gap between pathology and etiology as the experience of applying the taxonomic con-figurations in a "just" manner. "Justice" demands that a singular occurrence of a con-figuration be responded to with a unique symptomatic characteristic. Thus, an unconfigurable interpretation of an idiopathic condition is a leap from indeterminancy towards determinancy.

In logosophistic medicine, etiology takes on the structure of a marking according to which absence and impossibility can be made present and possible. Insofar as logosophistic diagnosis is motivated by such a promise, it escapes the traditional presence/absence dichotomy because a con-figuration is neither present nor absent. Therefore, an unconfigurable diagnosis will never definitively achieve "justice" in order to be considered perfectly right. Thence, a logosophistic diagnosis is unconfigurable, and in this way it is always deferred; however, every deference reveals a new facet of the Logos, and in this manner it builds a new etiologic stratum.

The terminology of logosophistic diagnosis

Logosophistic diagnosis makes use of a number of terms – some of which are coined or repurposed – that illustrate or follow its processes. Here are some of the most frequently used terms.

Disparity

Against the indeterminancy of symptoms, logosophistic diagnosis brings a concept called "disparity." This word is, on the re-constructive argument, properly neither a term nor a concept; it thereby names the non-coincidence of unconfigurability, both synchronically and diachronically. Because the resonance and conflict between determinancy and indeterminancy is difficult to convey in English, the term "disparity" is used for the lack of correspondence, inconsistency, or gap existent within a certain physiological system.

Outline

The idea of disparity also brings with it the idea of an outline. An outline is what a symptom both differs and defers from. It is the absent part of symptom's presence. In other words, through disparity, a symptom draws an outline, which is whatever is left over after everything present has been accounted for. The outline itself does not exist in itself because it is diffident. That is, in the act of presenting itself, the symptom becomes effaced. Because all symptoms viewed as present will necessarily draw outlines of other (absent) symptoms, the etiologic signifier can be neither wholly present nor wholly absent.

Idiopathanalogy

In logosophistic diagnosis, the term "idiopathanalogy" is appropriated to refer not just to systems of idiopathologic networking, but to all systems affected by a form of disparity that leads back to the hypothetical Logos (*ana-Logos*). A related term, called "archi-idiopathanalogy," refers to the positive side of idiopathanalogy, or disparity as an ultimate principle, rather than as a derivative of the *ana-Logos*. In other words, whereas the outline encompasses disparity, it is equally valid to view archi-idiopathanalogy as encompassing the outline, and therefore idiopathic pathology can be thought of as a form of outline.

Surplement, originary gap, and in-investigation

The coinage "surplement" is made up of the English "supplement" and the French "sur"; it is essential extra added to something complete in itself. Western medical thinking especially is characterized by the logic of over-supplementation, which is actually two apparently conflicting ideas. From one perspective, a supplement serves to enhance the presence of something which is already complete and self-sufficient. Thus, the outline is the surplement of disparity, unconfigurability is the surplement of itiology, and so on. But simultaneously, the idea of the surplement has within it the idea that a thing that has an

15

over-supplement cannot be truly complete in itself. If it were complete without the over-supplement, it should not need any supplement at all. The fact that a thing can be added- to make it even healthier or "whole" means that there is a lack – which we shall call from now on "originary gap." The surplement can fill that hole. The medical opening of this "hole" is called "in-investigation." From this perspective, the surplement does not enhance the presence of a disease, but rather underscores its absence.

Thus, what really happens during surplementation is that something appears from one perspective to be whole, complete, and self-sufficient, with the surplement acting as an external appendage. However, from another perspective, the surplement also fills a hole within the interior of the original Logos. Thus, the surplement represents an indeterminacy between interiority and exteriority, that is, between idiopathology and etiology.

Conclusion

Neo-pragmatist schools of medicine will be able to utilize logosophistic diagnosis not necessarily as a new methodology – although this is what logosophistic diagnosis ultimately is – but as something that *is already, all the time* occurring during the etiologic process. Logosophistic diagnosis is close to positing certain research protocols, methodologic gestures, and structural approaches which are intrinsic to all diagnoses, and thus close to positing an essential privileged contextualization of a medical condition. Thus, logosophistic diagnosis abandons the tendency to treat every disease as "about" the same old medical oppositions, healthy and unhealthy, positive and negative, sane and insane, intelligible unintelligible, subject and object, etc.

According to logosophistic medicine, in making the tacit assumption that all structures and diagnoses in traditional medicine are always and already present within all medical discourse, logosophistic diagnosis re-elevates differential diagnosis to a position at the *centre* of medical practice, a notion which neo-pragmatism in medicine seeks to eschew at all costs. This is an attempt to magnify the importance of a *medical generality*, at a time when specialization in medicine is more visible than ever. In addition, logosophistic medicine regards the attempts to privilege medical contextualization over other medical approaches, and to repeatedly prove the undecidedness of all forms of medical diagnosis.

Logosophistic diagnosis still accepts the validity of de-constructive differential etiologies, but view them as the result of a subjective interaction with a contextualized condition. Hence, in logosophistic medicine each diagnosis is one of many possible diagnoses, rather than the "excavation" of a condition buried within an organism. The "truth" of any single diagnosis is thus not privileged in that case, but open to further critical diagnosis.

D.D.

Medical Notes

The hybrid paramecia that secrete into the medium loci particles that kill other hybrid paramecia have a logonic structure identical with the mycoplasma, and that is why in many types of tumors they can unite transitional forms with their ancestors in a spectrum of strains held together by two or more disulfide bridges.

*

Although it is well known that the compensatory responses to alkalosis are fundamentally opposite to those occurring in acidosis, in respiratory acidosis the damage to the respiratory centre in the medulla oblongata causes, because of a decrease in the number of logons, an increase in extracellular fluid pH and a decrease in hydrogen ion concentration which also obstructs the passageways of the respiratory tract by interfering with the exchange of gases between the blood and the alveolar air.

*

In the so-called prophase, the initial stage of the first division of meiosis in which the chromosomes become visible and the homologous pairs of chromosomes undergo synopsis and crossing-over, the proliferation of logons causes the promyelocytes and all the other differentiated granulocytes in the marrow of a bone located in a pronator present an unstructured or ambiguous projection area in the form of a racemose cluster that amplifies harmful free radicals of oxygen formed during normal metabolic cell processes to oxygen and hydrogen peroxide, though without being a superoxide dismutase since its enzyme does not contain any metal.

*

The monoblast, since it is a mobile cell giving birth to the monocyte of the circulating blood, when overpopulated by monovalent generations of logons, produces an increase in potential difference across a hyperplastic tissue, and this often results in hyperoxia and scotomata.

*

The stratum corneum of the skin, when hyperkeratosed, presents a greater than typical proportion of logons in the cells which exhibit a higher than normal sensitivity to immunization because the antibodies in a patient suffering from hyperkeratosis are characterized by an uninhibited response to substructural stimuli.

*

Onset pneumonia is a severe atypical lobar pneumonia characterized, besides chill, fever, difficulty in breathing, cough, and blood-stained sputum, by motor and sensory nerve damage resulting in peculiarities of gait, impairment of vision, lassitude or extreme excitement, emaciation, and ultimately paralysis and death, unless controlled. I have called it *onset pneumonia* because of its sudden onset, which is in fact its chief characteristic.

*

The monoblast, since it is a motile cell giving birth to the monocyte of the circulating blood, when overpopulated by monovalent generations of logons, produces an increase in potential difference across a hyperplastic tissue, and this often results in a condition characterized by hyperoxia and scotomata.

*

The excitation of the lateral femoral cutaneous nerve in the patients with Lassa fever is a reflex condition caused by the hemorrhages under the skin which covers a region of the body starting with the lumbar plexus and ending with the anterior and lateral aspects of the thigh down to the knee.

*

The complete inhibition of other intact genetic crossovers generated by a surplus of logons in the vicinity of a chromosomal locus where a preceding crossover has occurred results in the prevention of typical growth and development of a virus in a suitable host because of the presence of another virus in the same host individual.

*

Endoplasmic fibroplasia must be associated with the conversion of the endocardium to fibroelastic tissue in most serious cases of congestive heart failure.

*

Besides abdominal distension, cyanosis, vasomotor collapse and irregular respiration, the gray syndrome, which affects especially premature infants, presents a damage of the lateral collateral ligaments of the knee resulting in lateral dislocation of the joint

*

The epiandrosterone occurring in most eneureses presents a logonic morphology which resembles that of the homogentisic acid formed as an intermediate in the metabolism of phenylalanine and tyrosine and found in the urine of those affected with alkaptonuria.

*

The increased production of watery liver bile results, besides the regular symptoms of hydrocholerosis, in cleft palate, flattened facies, and mild joint dislocations, which often makes it easy to be confused with Larsen's syndrome, especially when there is no increased secretion of bile solids.

*

The flexure limitations of the terminal phalanges of the fingers in most cases of severe arthritis may be related to a granuloma in one or more deep muscles of the ulnar side, especially the flexor digitorum superficialis.

*

The abuse of tartar emetic, the poisonous crystalline salt used as an expectorant in the treatment of amebiasis, results in one of the several types of glomerulonephritis that are characterized by proteinuria and hematuria and that often lead to renal failure.

*

The excessive deposition of hemosiderin occurring in most patients affected by endocardial fibroelastosis is due to a massive deficiency of iron. The lack of iron obstructs the transport of oxygen in the body, and this causes the progressive enlargement of the heart that is characterized by conversion of the endocardium to fibroelastic tissue.

*

Eosinophilic granuloma occurs in the Lesch-Nyhan syndrome because of a hypoplastic condition which affects the development of an organ. This organ remains below the normal size or in an immature state because of an acute deficiency of all logosophistic sources of hypoxanthine-guanine phosphoribosyltransferase.

*

Isosorbide dinitrate, the coronary vasodilator used in the treatment of angina pectoris, when administered to patients suffering from isovaleric academia, increases the seriousness of the acidosis which normally occurs in these patients, and causes isotonic jactitation, which is most of the time confused with a psychiatric disorder.

*

Granuloma annulare may be caused by a primitive form of infectious mononucleosis, an acute infectious disease associated with the Epstein-Barr virus. The fact that in granuloma annulare there are neither signs of lymphocytosis nor swellings of lymph nodes might be the result of the lack of innervated inotropism in most parts of the body, except for the feet, legs, hands and fingers, on which a benign chronic rash is present, this eruption being characterized by one or more flat spreading ring-like spots with lighter centres.

*

Among the classic symptoms of Piaget's disease, such as the inflammation of the nipple and areola, one may notice in some rare cases a fetid discharge from the nose similar to that of ozena, plus pachyonychia, an instance which makes us believe that the latter is not always a congenital disease.

*

The palindromic structure of the DNA denotes that it is evolutionarily old, because its double-stranded sequence in which the order of the nucleotides is the same on each side has a relatively less complex organization, despite the fact that the nucleotides run in opposite directions, which might be the result of logonic superswarming.

*

The asymmetrical or local enlargement and sclerotic changes in the long bones of one extremity noticed in melorheostosis may be accompanied by melena, an indication of bleeding in the upper part of the alimentary canal caused by the hardening of bone marrow.

*

The hypertrophy of the lower portion of the esophagus in megaesophagus may be related to the presence of megaloblasts in the blood, an anemia which may facilitate, if not properly treated, paraesophageal hernia, a severe condition in which the stomach herniates through the hiatus into the thorax.

*

The hyperanalgesic nature of some endorphins may cause some parts of the brain to epithalize and, because of this, the patient suffering from a surplus of endorphins may display epileptoid symptoms.

The conversion of the endocardium to fibroelastic tissue in endocardial fibroelastosis may be caused by an excessive amount of nisin, especially when absorbed from canned fruits and vegetables, or from cheese in which it is frequently used as a food preservative. This polypeptide antibiotic, which is produced by a bacterium of the genus *Streptococcus*, may lay a heavy influence on other conditions usually associated with congestive heart failure.

*

Somatization is not a psychological disorder. The fact that no organic or physiological explanation is found does not imply that psychological factors are involved. Somatization is a somatoform disorder characterized by multiple and recurrent physical *symptoms* (and not "complaints," as they are usually labeled) which are due to a general dysfunction of the organism, a *metastasis* of the entire soma without necessarily causing the death of the patient. The metastatic nature of the somatization disorder consists of the transfer of a disease from the original site of disease to another part of the body with development of a similar lesion in the new location, but without a disease-producing agency, such as cells or bacteria.

*

Mild symptoms of dyspraxia and dysphasia are visible in all patients with dysproteinemia.

*

Paraproteinemia may also be caused by a dysfunction of the parathyroid gland. The presence of abnormal serum globulins in the blood may be the result of an anomaly of the parathyroid gland; this small gland adjacent to the thyroid gland is composed of secretary epithelial cells lying in a stroma which is sometimes too poor in capillaries, and this causes the presence of the paraprotein in the blood.

*

The increase of the haptoglobins in the patients diagnosed with Hartnup disease affects the cerebellum, and subsequently creates various forms of muscular incoordination. The increase of haptoglobins is also related to aminoaciduria, which involves only monoamines having a single carboxyl group.

*

The lateral brachial cutaneous nerve is a continuation of the posterior branch of the axillary nerve that supplies the skin of the lateral aspect of the upper arm over the distal part of the deltoid muscle and the adjacent head of the triceps brachii. Any pressure on this nerve, which is due to a logonic surplus, will be radiated towards the lateral lemniscus, a band of nerve fibers that arises in the cochlear nuclei and terminates in the inferior colliculus and the lateral geniculate body of the opposite side of the brain. This explains why the somatic aspect of a damaged lateral brachial cutaneous nerve influences the mental of the patients suffering from this condition, and that is why they may display symptoms of severe incoordination and other neurological imbalances.

*

Polypnea accompanies frequently cystic fibrosis, but the rapid or panting respiration is not caused by mucus accumulation in the airways, as it is generally believed. This type of polypnea has a cytological cause, and a complete cytodiagnosis must be run on all patients affected by cystic fibrosis.

*

All cytopathogenic elements are directly or indirectly producing pathological changes in cells as a result of massive logonic accumulations, which create megalic logon surpluses.

*

Interferons, because of their low molecular weight, are responsible for most oxidation-reduction processes in the body, that means for almost all chemical reactions in which electrons are transferred from one molecule to another.

Most metabolic diseases are caused by deficient hexosaminidases, that is by a serious surplus of logons in the hydrolytic enzymes that catalyze the splitting off of a hexose from a ganglioside.

*

Granulocytosis may be related to the damage of the gray commissure, the transverse band of gray matter in the spinal cord appearing in sections as the transverse bar of the H-shaped mass of gray matter.

*

Mononucleosis, because of its abnormal increase of mononuclear leukocytes in the blood, which causes various degrees of logonic accumulation, also produces the inflammation of the mastoid cells, this being the most severe form of mastoiditis.

*

When the erythrocyte sedimentation rate in anemia is not above normal, antibodies to double-stranded DNA occur, resulting in damage to normal tissue and inflamed muscles.

*

A special type of myxoma-a soft tumour made up of gelatinous connective tissue-is highly myotonic. The symptoms of myotonic myxoma include tonic spasms of one or more muscles and delay in the ability to relax muscles after forceful contraction.

*

Moebius syndrome might be related to a bacterium of the genus *Streptomyces* (*S. cinnamoensis*), which is associated with other neurological disorders caused by protozoal agents.

*

Among the symptoms regularly associated with perihepatitis, the inflammation of the peritoneal capsule of the liver, one must include the inflammation of the infraorbital nerve, a branch of the maxillary nerve that divides into branches distributed to the skin of the upper part of the cheek, the upper lip, and the lower eyelid. This symptom might be related to the infraorbital vein, which drains the inferior structures of the orbit and the adjacent area of the face, and which empties into the pterygoid plexus.

*

Xerosis and xerostomia are present in Duchenne's disease because of keratomalacia, a dry thickened lustreless condition of the eyeball, which is also correlated with a severe systemic deficiency of vitamin A.

*

Lentigines develop in an identical way with leiomyomata because they both fail to stop the growth of the same amount of logons. It still has to be explained why the former turn into melanotic spots in the skin in which the formation of pigment is unrelated to exposure to sunlight, while the latter develop into smooth muscle fibers.

*

The formation of blood granulocytes, typically in the bone marrow, in the condition called granulopoiesis also affects the cells of the epithelial lining of a graafian follicle or its follicular precursor.

*

Lomustine, the antineoplastic drug used in the treatment of brain tumours and Hodgkin's disease, may also be effective in the treatment of Graves' disease because of its impact on the so-called long-acting thyroid stimulator, a protein that occurs in the plasma of patients with hyperthyroidism.

*

A great number of cryptogenic infections present symptoms that include abnormal quantities of cryoglobulins, that means proteins similar to gamma globulins which precipitate usually in the cold from blood serum, especially in pathological conditions. This implies that most cryptogenic infections have the same mechanism of disease as multiple myeloma or cryoglobulinemia.

*

We shall call *hydropic cystic fibrosis* a particular case of cystic fibrosis in which the deficiency of pancreatic enzymes is characterized by swelling and taking up fluid.

*

Longus capitis is a muscle of either side of the front and upper portion of the neck that arises from the third to the sixth cervical vertebrae. We have noticed that its role in lordosis is considerable, since an atrophy of the longus capitis increases the forward curvature of the lumbar and cervical regions of the spinal column.

*

Myonecrosis may be noticed in several cases of myxedema, a severe form of hypothyroidism characterized by firm inelastic edema, dry skin and hair, and loss of mental and physical vigour. The muscle necrosis in myxedema is related to the abnormal excess accumulation of serous fluid in connective tissues.

The inflammatory distension of one or both fallopian tubes with fluid in hydrosalpinx leads to an abnormal surplus of hydrochloric acid, an aqueous solution of hydrogen chloride HCl that is a strong corrosive irritating acid and is present in dilute form in gastric juice.

*

Coronaviruses impact both the sense of seeing and that of hearing because they damage the corpora quadrigemina, which is made up of two pairs of colliculi on the dorsal surface of the midbrain composed of white matter externally and grey matter within, the superior pair containing correlation centres for optic reflexes and the inferior pair containing correlation centres for auditory reflexes.

*

Commissurotomy of the common peroneal nerve-the smaller of the branches into which the sciatic nerve divides, passing outward and downward from the popliteal space-gives birth to what we have called *chronotropic clonus*, consecutive series of alternating contractions and partial relaxations of the peroneal muscle which modify the rate of the heartbeat.

*

A symptom of aggravated pulmonary alveolar proteinosis-a chronic disease of the lungs characterized by the filling of the alveoli with protenaiceous material-is pulmonary stenosis, an abnormal narrowing of the orifice between the pulmonary artery and the right ventricle.

*

Laminin, a glycoprotein that is a component of connective tissue basement membrane and that promotes cell adhesion, has a lancinating structure in the Langhans giant cell or in any cells found in the lesions of granulomatous conditions and containing a number of peripheral nuclei arranged in a circle or in the shape of a horseshoe.

*

Gasterophilus is a genus of botflies that can infest not only horses, but humans as well. When that happens, it creates gastrinomata, neoplasms occurring in the pancreas or the wall of the duodenum, and producing excessive amounts of gastrin, a hormone that is secreted by the gastric mucosa and that induces secretion of gastric juice.

*

The Heinz body is a cellular inclusion in a red blood cell that consists of damaged aggregated hemoglobin. Its implications in haemolytic anemia is due to the fact that it blocks the normal functioning of the helper T cell, and thus the latter stops participating in an immune response by recognizing foreign antigens and secreting lymphokines to activate T cell and B cell proliferation, that usually carries CD4 molecular markers on its cell surface.

*

Penetrance, the proportion of individuals of a particular genotype that expresses its phenotypic effect in a given environment, has a fundamentally logonic structure since it follows all the physical and chemical laws of logonic biology.

*

Hepatogenic logons become hemotoxic when any of the epithelial parenchymatous cells of the liver, which are also known as hepatocytes, modify their polygonal morphology.

*

Finasteride, the antineoplastic drug used especially to shrink an enlarged prostate gland, may cause fibrotitis, a muscular condition commonly accompanied by the formation of painful subcutaneous modules, and that is due to the highly inhibited conversion of testosterone to dihydrotestosterone.

*

The prohormonic status of prolactins in some cases of defective lactation processes may be related to a dysfunctional projection area in the cortex, and this is due to its connection, through projection fibres, with the subcortical cantres that in turn are linked with peripheral sense and motor organs.

*

Aplastic corneas should be treated as a corneoscleral condition, since the faulty development of the cornea affects both the cornea and the sclera.

*

Cytopenia, a deficiency of cellular elements of the blood may be responsible for the cellular enlargement and formation of eosinophilic inclusion bodies caused by cytomegaloviruses, which are the major source of cytomegalic inclusion disease, a severe condition especially of newborns, affecting the salivary glands, brain, kidneys, liver, and lungs.

*

The breakdown of hemoglobin in hemosiderin is caused not only by disturbances of iron metabolism (as in hemochromatosis, hemosiderosis, or some anemias), but also by an insufficient number of hemostatic agents or a surplus of heparin, especially when the latter is administered parenterally as the sodium salt in vascular surgery and in the treatment of postoperative thrombosis and embolism.

*

Cytochrome oxidase is an iron-porphyrin enzyme with a major role in cell respiration. Its ability to catalyze the oxidation of reduced cytochrome c in the presence of oxygen is based on the dissolution or disintegration of logonic structures.

*

Gastrinoma is a neoplasm that involves blood vessels, usually occurring in the pancreas or the wall of the duodenum. The excessive amounts of gastrin produced by the gastrinoma may affect the mucous membrane of the stomach and finally the gastric artery, that means a branch of the celiac artery that passes to the cardiac end of the stomach and along the lesser curvature.

*

Selection, the natural or artificial process that results or tends to result in the survival or propagation of some individuals but of no others, is dependent on self-tolerance and self-assembly: the physiological state that exists in a developing organism, when its immune system has proceeded far enough in the process of self-recognition to lose the capacity to attack and destroy its own bodily constituents, is directly related to the process by which a complex macromolecule or a supramolecular system spontaneously assembles itself from its own bodily constituents.

*

Metacresol is an isomer of cresol that has, besides antiseptic properties, the metachronous capability to metabolize, which is virtually unknown to specialists.

*

Penicillamine is an amino acid that is obtained from penicillins and is used especially to treat cystinuria and metal poisoning, as by copper and lead. Nobody else has noticed so far that penicillamine is also a mild stimulant of the central nervous system.

*

There is a hypothetical class of logonic antibodies IgL of infinitesimal molecular weight that function in allergic reactions and are active against bacteria, viruses, and proteins foreign to the body. These logonic antibodies appear early in the immune response and then are replaced by antibodies of high molecular weight (IgM).

*

Laryngismus stridulus, the sudden spasm of the larynx that occurs in children may be related to satellitosis, a condition characterized by a grouping of satellite cells around ganglion cells in the brain.

*

Coronaviruses, besides respiratory symptoms, also cause symptoms of cor bovinum and may produce a dysfunctional Cori process, the cycle in which carbohydrate metabolism consists of the conversion of glycogen to the lactic acid in the muscle, the diffusion of the lactic acid into the bloodstream, and the breakdown of liver glycogen to glucose.

*

Moh's technique can be successfully used in cases of molluscum contagiosum when the removal of the nodules cannot be performed without a deep excision.

*

Chemopalidectomy, the destruction of globus pallidus (the median portion of the lentiform nucleus), may be effective in several types of chemodectoma, a tumour that affects tissue.

*

Cyproterone, the synthetic steroid used in the form of its acetate to inhibit androgenic secretions, can also be used in the treatment of asthma because it acts antagonistically to histamine and serotonin.

*

Opportunistic logons are logons that are usually harmless but can become pathogenic when the organism's resistance to disease is impaired.

*

Logonic gliomatosis is a glioma (tumour arising from neuroglia) with a rather concentrated than diffuse proliferation of glial cells and with a single focus rather than multiple foci, caused by opportunistic logons (see the previous entry).

*

Devascularization occurs in some cases of endocardial fibroelastosis, because the conversion of the endocardium to fibroelastic tissue occasionally produces an obstruction or destruction of blood vessels, and this causes a loss of the blood supply to a certain bodily part.

*

The upstream of uracil, the pyrimidine base that codes genetic information in the polynucleotide chain of RNA, has a hydroxyl group attached to the position labelled 5' in the terminal nucleotide, which is responsible, among other things, for the hydrolysis of urea into ammonia and carbon dioxide through the urease, an enzyme that actually catalyzes the above-mentioned hydrolysis.

*

The complement-fixation test is a diagnostic test for the presence of a certain antibody in the serum of a patient that involves a particular inactivation of the complement in the serum, addition of carefully measured amounts of the antigen for which the antibody is specific and of extraneous complement, and indication of the presence/absence of complement fixation by the addition of a certain type of indicator. The risk of obtaining erroneous results in the complement-fixation test is abnormally high because of the union of an antibody and the antigen for which it is specific that occurs when the complement is added to a mixture of such an antibody and antigen can be compromised by the latter's absence of complement fixation in the enzyme's logonic structure, which is responsible for its inability to stimulate an immune response.

*

Disorders of the dictyosome-any of the membranous or vesicular structures making up the Golgi apparatus-often cause a dicrotic notch, that means a secondary upstroke in the descending part of a pulse tracing, corresponding to the transient increase in aortic pressure upon closure of the aortic valve.

Interleukin-2 is a compound of low molecular weight produced by antigen-stimulated helper T cells in the presence of interleukin-1 that induces proliferation of immune cells in the form of T cells and B cells. Interleukin-2 may be helpful in the treatment of disorders related to defective interkinesis processes, since the surplus of T cells and B cells regulates the first and second meiotic divisions.

Fibronectins, the glycoproteins of cell surfaces, blood plasma, and connective tissue, promote not only cellular adhesion and migration, but also metamesoblastic differentiation.

The visual process of lateral inhibition consists of the firing of a retinal cell inhibiting the firing of surrounding retinal cells, and is related to the preference in use of homologous parts on one lateral half of the body over those on the other. This dominance in function of one of a pair of lateral homologous parts is held to enhance the perception of areas of contrast in lateral inhibition.

The presence of large numbers of monocytes in the circulating blood, in the case of monocytic leukemia, deaminates monoamines oxidatively and functions in the nervous system by breaking down monoamine neurotransmitters oxidatively. That is why, patients suffering from monocytic leukemia should also be prescribed monoamine oxidase inhibitors, which increase the concentration of monoamines by inhabiting the action of monoamine oxidase.

*

The blood of the patients affected by Chagas' disease, a tropical American disease that is caused by a flagellate of the genus *Trypanosama* (*T cruzi*), loses its ability to secret the cerebrospinal fluid into the lateral ventricles of the brain, and that causes the pressure within the brain and spinal cord lose its uniformity.

*

The pathologic dissolution or disintegration of cells encountered in the several types of cytolysis is caused by successive logonic inclusions and capacitations.

*

Inhibiting an enzyme catalyzing conversion of testosterone to dihydrotestosterone by such antineoplastic drugs as finasteride may also result in disruptive fibronectins conducive to various types of fibrous ankylosis.

*

The dense logonic accumulation occurring in hemocytoblasts may be due to hemopneumothorax, an accumulation of blood and air in the pleural cavity.

*

Hemochromatosis – the metabolic disorder that is characterized by deposition of iron-containing pigments in the tissues and frequently by diabetes and weakness – is sometimes related to the decreased concentration of cells and solids in the blood resulting from gain of fluid from the tissues, thus causing it to be classified as a particular type of hemodilution, which we shall call *hemochromatotic dilution*.

*

All muscular dystrophies of the Duchenne type must be also related to a dysfunctional claustrum, one of the four basal ganglia in each cerebral hemisphere consisting of a thin lamina of grey matter between the lentiform nucleus and the insula.

*

The cytoarchitecture of cystine – a crystalline amino acid that is widespread in such proteins as keratins and is a major metabolic sulphur source – presents a dense cellular makeup consisting of massive logonic accumulations.

*

The M substance, which is a protein that is an antigen tending to occur on the surface of beta-hemolytic bacteria belonging to the genus *Streptococcus* and that is placed in a particular group (Lancefield group A), is also responsible for the Kawasaki disease, especially when it presents erythema of the conjunctivae and of the mucous membranes of the upper respiratory tract, erythema and edema of the hands and feet, and cervical lymphadenopathy.

*

Lathyrogen, the compound that tends to cause lathyrism and inhibit the formation of links between chains of collagen, although it normally affects the legs, may also cause the paralysis of the latissimus dorsi, a broad flat superficial muscle of the lower part of the back that extends, adducts, and rotates the arm medially and draws the shoulder downward and backward.

*

Ruptured graafian follicles of the corpus hemorrhagicum kind, since they contain a blood clot that is absorbed as the cells lining the follicle form the corpus luteum, represent a natural progressively continuing logonic accumulation resultant in the creation of the progesterone, the female steroid sex hormone that is secreted by the corpus luteum to prepare the endometrium for implementation and later by the placenta during pregnancy to prevent rejection of the developing embryo or fetus.

*

Exflagellation, the formation of microgametes in sporozoans by extrusion of nuclear material into peripheral processes, must become the study object of exfoliate cytology because the cells shed from the body surfaces present the same mechanisms of logonic accumulation as the processes that resemble flagella.

*

Papaverine, the crystalline alkaloid that is used in the form of its hydrochloride mainly as an antispasmodic, is also efficient in the treatment of paracholera, especially in combination with parachlorophenol, a chlorinated phenol chiefly used as a germicide.

*

Gastroduodenal artery, the artery that arises from the hepatic artery, contains a small number of logonic accumulations because it divides to form the right gastroepiploic artery and a branch supplying the duodenum and pancreas, all these having their own logonic accumulations. Due to the small number of logons contained in its cells, the gastroduodenal artery is prone to various diseases.

*

Researchers have to check the quantity of logonic accumulations in the massa intermedia, an apparently functionless mass of grey matter in the midline of the third ventricle that is found in many but not only human brains and is formed when the surfaces of the thalami protruding inward from opposite sides of the third ventricle make contact and fuse. What they could find might explain a possible purpose of this so-called futile mass of grey matter.

*

Elliptocytis, the trait manifested by the presence in the blood of red cells which are oval in shape with rounded ends, may degenerate into thrombocytopenic purpura, a condition that is characterized by bleeding into the skin with the production of petechiae or ecchymoses if the reduction of the logonic accumulation caused by elliptocytosis also produces a reduction in circulating blood platelets.

*

The hepatization of the lungs occurring in some cases of hepatoma, the malignant tumour of the liver, consists in the conversion of the tissue of the lungs into a substance resembling liver tissue. This transformation is related to the sympathetic characteristic of the logons.

*

Exsiccosis, the insufficient intake of fluids, encountered in most cases of bladder exstrophy – a congenital malformation of the bladder in which the normally internal mucosa of the organ lies exposed on the abdominal wall – results in severe bodily dehydration that can damage the exopeptidase, the enzyme that hydrolyse peptide bonds formed by the terminal amino acids of peptide chains.

*

Exsanguination, the process of draining or losing blood, can be stopped with expanders (a colloidal substance, such as dextran, of high molecular weight used as blood or plasma substitute for increasing the blood volume) with a dense logonic accumulation.

(Toronto, Ontario, 2005)